Cambridge Elements ≡

Elements of Paleontology
edited by
Colin D. Sumrall
University of Tennessee

COMPUTATIONAL FLUID DYNAMICS AND ITS APPLICATIONS IN ECHINODERM PALEOBIOLOGY

Imran A. Rahman

CAMBRIDGE
UNIVERSITY PRESS

CAMBRIDGE
UNIVERSITY PRESS

University Printing House, Cambridge CB2 8BS, United Kingdom

One Liberty Plaza, 20th Floor, New York, NY 10006, USA

477 Williamstown Road, Port Melbourne, VIC 3207, Australia

314–321, 3rd Floor, Plot 3, Splendor Forum, Jasola District Centre, New Delhi – 110025, India

79 Anson Road, #06–04/06, Singapore 079906

Cambridge University Press is part of the University of Cambridge.

It furthers the University's mission by disseminating knowledge in the pursuit of education, learning, and research at the highest international levels of excellence.

www.cambridge.org
Information on this title: www.cambridge.org/9781108810029
DOI: 10.1017/9781108893473

First published 2020

A catalogue record for this publication is available from the British Library.

ISBN 978-1-108-81002-9 Paperback
ISSN 2517-780X (online)
ISSN 2517-7796 (print)

Computational fluid dynamics and its applications in echinoderm paleobiology

Elements of Paleontology

DOI: 10.1017/9781108893473
First published online: November 2020

Imran A. Rahman

Author for correspondence: Imran A. Rahman, imran.rahman@oum.ox.ac.uk

Abstract: Computational fluid dynamics (CFD), which involves using computers to simulate fluid flow, is emerging as a powerful approach for elucidating the paleobiology of ancient organisms. Here, Imran A. Rahman describes its applications for studying fossil echinoderms. When properly configured, CFD simulations can be used to test functional hypotheses in extinct species, informing on aspects such as feeding and stability. They also show great promise for addressing ecological questions related to the interaction between organisms and their environment. Computational fluid dynamics has the potential to become an important tool in echinoderm paleobiology over the coming years.

Keywords: computational fluid dynamics, echinoderms, paleobiology, function, ecology

ISBNs: 9781108810029 (PB), 9781108893473 (OC)
ISSNs: 2517-780X (online), 2517-7796 (print)

Contents

1 Introduction

Echinoderms exhibit a plethora of morphological and behavioural adaptations to life in moving fluids. For example, the slot-like holes (lunules) that pass through the test in many sand dollars serve to reduce lift and thereby increase resistance to dislodgement (Telford, 1983), while at the same time drawing food-laden water up from the substrate (Alexander & Ghiold, 1980). Moreover, stalked crinoids bend the proximal-most part of the stalk and crown down-current, with the arms arranged into a fan in order to improve particle capture during suspension feeding (Macurda & Meyer, 1974; Baumiller, 2008). Laboratory experiments and field observations of living organisms enable us to better understand how extant echinoderms interact with fluid flows (e.g. Macurda & Meyer, 1974; Alexander & Ghiold, 1980; Telford, 1983; Messing et al., 1988; Baumiller et al., 1991; Loo et al., 1996; Thompson et al., 2005; Holtz & MacDonald, 2009; Cohen-Rengifo et al., 2018), but for fossil taxa, especially those lacking a modern analogue, methods of investigation are more limited. Flume studies have been used to explore the feeding and hydrodynamics of extinct echinoderms based on physical models of fossil organisms, both historically (e.g. Welch, 1978; Baumiller & Plotnick, 1989; Riddle, 1989; Parsley, 1990; Friedrich, 1993; Daley, 1996) and more recently (e.g. Huynh et al., 2015; Parsley, 2015). However, given the increasing availability of three-dimensional, digital models of fossil echinoderms (e.g. Rahman & Zamora, 2009; Zamora et al., 2012; Zamora & Smith, 2012; Waters et al., 2015; Briggs et al., 2017; Clark et al., 2017; Reich et al., 2017; Bauer et al., 2019; Rahman et al., 2019; Saulsbury & Zamora, 2019), virtual modelling approaches have great potential for analysing the paleobiology of extinct forms.

One of the most promising approaches for interrogating function and ecology in ancient organisms is computational fluid dynamics, or CFD. This is a tool for simulating flows of fluids, such as water or air. Computers solve the governing equations that describe fluid motions and their interactions with boundaries. In this way, fluid flows can be simulated for digital models of solid objects. Computational fluid dynamics is routinely used in engineering to analyse design and optimize performance for structures and machines. Furthermore, in the past ten to fifteen years, it has become increasingly important in paleontology (Rahman, 2017). Among the first to apply CFD to fossils were Rigby and Tabor (2006), who simulated water flow around digital models of graptolites. Shiino and colleagues subsequently applied the technique to brachiopods (Shiino et al., 2009; Shiino & Kuwazuru, 2010, 2011) and trilobites (Shiino et al., 2012, 2014). More recently, CFD has been used to study extinct vertebrates (Bourke et al., 2014, 2018; Kogan et al., 2015; Liu et al., 2015; Wroe et al., 2018; Dec, 2019; Gutarra et al., 2019; Troelsen et al., 2019), Ediacaran organisms (Rahman et al., 2015a; Darroch et al., 2017; Gibson et al., 2019), ammonoids (Hebdon et al., 2020) and fossil echinoderms (Rahman et al., 2015b, 2020; Dynowski et al., 2016; Waters et al., 2017).

In this Element, I introduce some basic principles of fluid dynamics and describe the key steps in a paleontological CFD study. I also discuss the applications of CFD to extinct echinoderms, highlighting recent work on cinctans (Rahman et al., 2015b), stalked crinoids (Dynowski et al., 2016) and blastoids (Waters et al., 2017). I end by considering possible future directions in this area, including new avenues of research that could improve our understanding of echinoderm paleobiology.

2 Fluid Dynamics

The discipline of fluid mechanics is the branch of physics that deals with fluids and the forces acting on them. It can be further subdivided into fluid statics and fluid dynamics, the latter of which is concerned with fluid flows and thus relevant for understanding the interaction between living organisms and moving fluids. Three conservation laws that must be satisfied in fluid dynamics are the conservation of mass, the conservation of momentum and the conservation of energy. Additionally, it is assumed that fluids can be treated as continuous substances (the continuum assumption), rather than being composed of discrete molecules. Applying these laws to the volume through which fluid will flow, by expressing them in terms of mathematical equations, allows us to calculate properties such as flow velocity and pressure as functions of space and time, thereby solving problems in fluid dynamics.

Water is treated as an incompressible Newtonian fluid, with density and viscosity assumed to be constant. The fluid velocity is zero at all solid boundaries (no-slip condition) and increases with distance from the boundary, giving a velocity gradient (the boundary layer). Where the viscous forces are relatively large, the fluid flows orderly in parallel layers, with little or no mixing (i.e. laminar flow). In contrast, where the inertial forces dominate, fluid flow is characterized by chaotic motion and the formation of unsteady vortices (i.e. turbulent flow). The dimensionless Reynolds number (Re) describes the ratio of inertial to viscous forces, and is defined as:

$$\text{Re} = \frac{\rho U L}{\mu}$$

where ρ is the density of the fluid, U is the characteristic velocity, L is the characteristic dimension and μ is the dynamic viscosity of the fluid. For flow through closed conduits (e.g. pipes), the characteristic dimension is typically the diameter of the conduit, whereas for flow around objects, it is usually the width or length of the object. Low Re indicates the flow is mostly laminar, while high Re is indicative of predominantly turbulent flow; however, the critical Reynolds numbers over which flow transitions from laminar to turbulent will vary depending on the geometry. Flows with geometrically similar objects (i.e. scaled versions of the same shape in the same orientation) and the same Re are said to have dynamic similarity (assuming the Womersley number, which describes pulsatile flow frequency, is also constant), meaning the fluid flows will be identical.

Drag is the force that acts opposite to the relative motion of an object in fluid (i.e. parallel to the flow direction). It is dependent on the properties of the fluid and the geometry of the object. The dimensionless drag coefficient (C_D) relates the drag force to the fluid density, velocity and object geometry, and is defined as:

$$C_D = \frac{2F_D}{\rho U^2 A}$$

where F_D is the drag force exerted by the fluid, ρ is the density of the fluid, U is the characteristic velocity and A is the characteristic area (commonly the projected frontal area, wetted surface area or total surface area of the object). C_D can be used to compare the performance of different geometries at the same Re, assuming the boundary conditions are consistent.

Lift is the force that acts perpendicular to the flow direction. Similar to drag, it varies with the fluid properties and object geometry. The lift coefficient (C_L) is defined as:

$$C_L = \frac{2F_L}{\rho U^2 A}$$

where F_L is the lift force exerted by the fluid, ρ is the density of the fluid, U is the characteristic velocity and A is the characteristic area (typically the plan area).

The governing equations of fluid flow include the Navier–Stokes equations, which describe the motion of the fluid, and the continuity equation, which represents the conservation of mass. These equations can be simplified to make them easier to solve. Nevertheless, for all but the simplest problems, the equations must be solved numerically on a computer. This can be done by splitting up the flow domain into smaller cells, with the governing equations discretized and solved in each of these cells. This approach is termed *computational fluid dynamics*.

3 Steps in Computational Fluid Dynamics

Computational fluid dynamics dates back to the 1950s and 1960s, when the first computer simulations of fluid flows were undertaken (e.g. Evans & Harlow, 1957; Harlow & Welch, 1965; Hess & Smith, 1967). It has subsequently been used to address problems in a wide range of subjects, including paleontology (Rahman, 2017). In many cases, the same set of steps, outlined in this section, is followed.

The first step in a paleontological CFD study is to construct a digital model of the organism of interest. For an increasing number of extinct groups, three-dimensional virtual reconstructions of fossil specimens already exist. However, additional work to digitally correct taphonomic distortion and restore the original morphology of the organism (e.g. Lautenschlager, 2016) might be required. An alternative approach is to create digital models through box or NURBS modelling. This allows models to be constructed for taxa where complete, three-dimensionally preserved fossil specimens are not available, but is more subjective than tomographic or surface-based methods (Rahman & Lautenschlager, 2017).

In CFD simulations of flow around an organism, the computational domain surrounding the modelled organism must also be created. For three-dimensional models, this will typically consist of a cuboid or cylinder. The domain should be large enough to ensure full development of the flow around the model. A domain that extends three times the length of the model upstream, ten times the length of the model downstream and five times the size of the model in all other directions can be taken as a starting point. However, the optimal domain size will vary on a case-by-case basis, and sensitivity analyses should be undertaken to establish the most appropriate size (see later in this section).

Next, the domain is divided into a number of discrete cells (the mesh) (Figure 1A). The mesh is commonly made up of tetrahedral or hexahedral elements, with layers of prismatic elements at the fluid–solid interface to model the boundary layer. Increasing the number of mesh elements can improve the accuracy of the simulation, but will increase the memory require-ments and computation time. Similar to the domain size, sensitivity analyses should be carried out to determine the optimal mesh size.

The material properties of the fluid, such as density and viscosity, must be specified. The flow model is then selected, which establishes the governing equations that will need to be solved in the simulation. The choice of model depends on the flow regime, which is indicated by the Reynolds number (see Section 2). Laminar flow can be described by the Navier–Stokes equations, but turbulent flows are more complex and so time-averaged equations of fluid

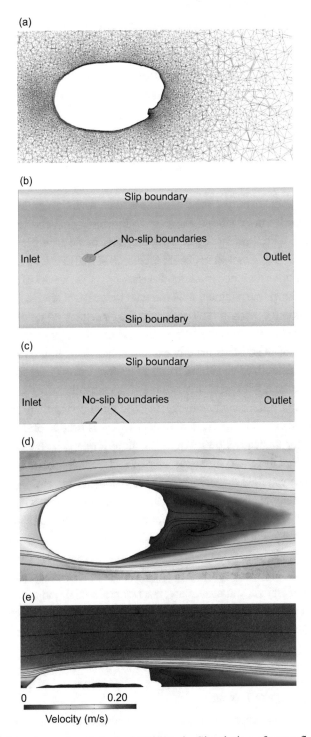

Figure 1 Steps in a paleontological CFD study. Simulation of water flow around a model of the cinctan *Protocinctus mansillaensis*. A: Two-dimensional plot

motion (the Reynolds-averaged Navier–Stokes, or RANS, equations) and a turbulence closure model are generally used. Various turbulence models exist, with the k-ε and shear stress transport (SST) models the most widely used in paleontological CFD analyses.

Following selection of the flow model, boundary conditions representing the flow variables are specified (Figures 1B, C). These include an inlet describing how flow enters the domain and an outlet that defines how it exits. Commonly, a velocity inlet and zero pressure outlet are used. For simulations of flow around a stationary organism, the inlet conditions will be informed by the current velocity, which can be inferred based on sedimentological characteristics (e.g. Stow et al., 2009) and/or direct measurements made in analogous modern environments (e.g. Emelyanov, 2005; Siedler et al., 2013). A no-slip boundary condition is often assigned to all solid surfaces (e.g. the model of the organism and the seafloor), meaning the fluid has zero velocity relative to the boundary. A slip boundary condition can be used for the remaining edges of the domain, allowing the flow to pass along these boundaries without friction.

When all of these steps have been completed, fluid flow is simulated by solving the discretized equations. This can be done using direct solvers, which are computationally expensive, or (more frequently) iterative solvers, which use less memory and can therefore reduce computation time. For steady flows, which do not vary temporally, a stationary solver is used to obtain a solution. However, for unsteady flows, where the flow properties change over time, a computationally much more expensive time-dependent solver must be used.

Computational fluid dynamics results are visualized and analysed in various ways. For example, plots of velocity magnitude supplemented with streamlines can be produced to study patterns of fluid flow around the modelled organism (Figures 1D, E). Furthermore, drag and lift forces and their coefficients (see Section 2) can be calculated to evaluate the forces exerted by the fluid on the organism. Testing the sensitivity of these results to simulation parameters, in particular the domain and mesh sizes, is a key part of the study. The size of the domain should be varied to determine the smallest possible domain that produces results matching theoretical expectations of flow development. Similarly,

---

Caption for Figure 1 (cont.)

(horizontal cross-section) of the mesh. B, C: Computational domain (top-down and side-on views, respectively) showing boundary conditions. D, E: Two-dimensional plots (horizontal and vertical cross-sections, respectively) of flow velocity with streamlines. Direction of ambient flow from left to right.

mesh independence must be established by undertaking simulations with different mesh sizes and identifying the coarsest mesh that did not substantially alter the results.

4 Examples in Echinoderm Paleobiology

Prior to CFD becoming part of the paleontologist's toolkit, studies of the hydrodynamics of fossil echinoderms relied primarily on experiments in flume tanks. Crinoids were an initial focus of this work. Welch (1978) introduced life-size models of the Carboniferous camerate *Pterotocrinus* into a flume, performing experiments with models at three different orientations to the current. The results demonstrated that redirection of water flow to the filtration fan was strongest when the models were orientated with the fan perpendicular to the current direction and the ambulacral side downcurrent (Figure 2A), similar to the feeding posture of extant stalked crinoids. Building on this, Baumiller and Plotnick (1989) carried out experiments for models of *Pterotocrinus* with and without the wing-like tegminal appendages. This showed that the models with wing plates rotated into a position with the ambulacral side of the fan pointed downcurrent. Riddle (1989) placed fossil specimens and models of crinoid columns in a recirculating flow tank. He found that the models with helically twisted columns, as seen in some platycrinitids, deflected water strongly upwards (Figure 2B), towards where the filtration apparatus would have been located in the living animal. Baumiller (1990) conducted experiments using models of batocrinids with and without an anal tube. This revealed that the models with an anal tube reduced flow from the anus to the feeding appendages.

This experimental approach was also extended to other extinct echinoderm groups. Parsley (1990) studied flow around models of the diploporitan *Aristocystites*, orientated with the aboral end of the theca facing into the current. He was able to show that vortices were generated downcurrent of the oral end, transporting particles to the ambulacra. A similar pattern was documented for cinctans by Friedrich (1993), who found that the models orientated with the stele facing upcurrent created back eddies that brought particles towards the mouth and marginal grooves. Daley (1996) undertook flume experiments using a model of the solute *Coleicarpus sprinklei*, which revealed that turbulence was created around the single feeding appendage when the model was positioned close to the substrate. Huynh and colleagues (2015) investigated fluid flow in blastoid respiratory structures using a 72x scale model of part of the hydrospire of *Pentremites rusticus*. They observed that there was no mixing of water taken in through incurrent hydrospire pores within the associated hydrospire folds

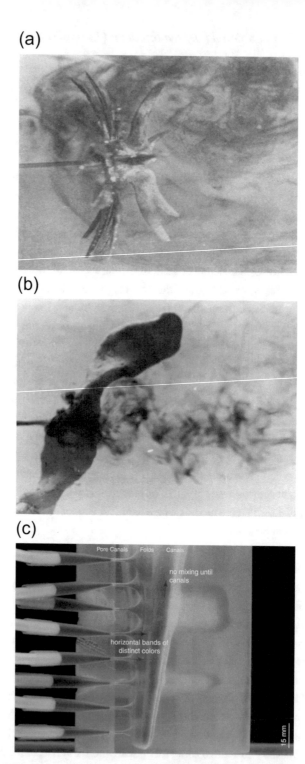

Figure 2 Flume experiments of fossil echinoderms. A: Water flow around a model of the crinoid *Pterotocrinus*. Modified from Welch (1978, fig. 3A),

(Figure 2C). Most recently, Parsley (2015) examined flow around models of the eocrinoids *Globoeocrinus*, *Guizhoueocrinus* and *Sinoeocrinus* in a flume tank. The results showed that the modelled brachioles were distally bent downcurrent by the flow, forming an open fan.

Computational fluid dynamics has considerable potential for expanding on this existing base of experimental work, enabling analyses of a wider range of model geometries and flow conditions. However, to date, only a few studies have applied the technique to fossil echinoderms. Rahman and colleagues (2015b) used the approach to explore the hydrodynamics of feeding in cinctans. Owing to their highly unusual body plan, which has no analogue among extant taxa, it is debated whether cinctans were passive suspension feeders (e.g. Parsley, 1999; David et al., 2000) or active filter feeders (e.g. Smith, 2005; Zamora & Smith, 2008). Computational fluid dynamics was used to test between these competing hypotheses. A digital model of the cinctan *Protocinctus mansillaensis*, created from an X-ray microtomography scan of the holotype (Rahman & Zamora, 2009), was placed in a virtual flume tank. Computer simulations of water flow were performed at inlet velocities of 0.05, 0.1 and 0.2 m/s, chosen to represent typical current velocities in the offshore environments inhabited by *Protocinctus* (Álvaro & Vennin, 1997). Models were positioned at different orientations and burial depths, and passive and active feeding scenarios were simulated by varying the boundary conditions at the main body openings. The results showed that in all cases the models positioned with the mouth downcurrent and the ventral swelling buried generated the least drag and lift, suggesting this position was optimal for enhancing stability. Moreover, in this orientation there was very little flow to the mouth and associated marginal groove in the simulations of passive feeding, demonstrating that such a feeding strategy would have been an ineffective way of obtaining nutrients. Conversely, there was strong flow to the mouth in the simulations of active feeding (Figure 3A), indicating that this was the most probable feeding mode and supporting previous interpretations of cinctans as pharyngeal filter feeders (e.g. Smith, 2005; Zamora & Smith, 2008).

Caption for Figure 2 (cont.)

reproduced with permission from Cambridge University Press. B: Water flow around a model of the column of the crinoid *Platycrinites*. Modified from Riddle (1989, fig. 1.3), reproduced with permission from Cambridge University Press. C: Water flow within a model of part of the hydrospire of the blastoid *Pentremites rusticus*. Modified from Huynh and colleagues (2015, fig. 4), reproduced under a CC BY-NC-SA 4.0 license.

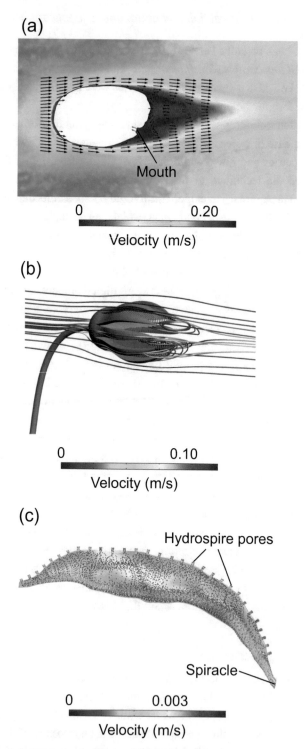

(a)

Mouth

0   0.20

Velocity (m/s)

(b)

0   0.10

Velocity (m/s)

(c)

Hydrospire pores

Spiracle

0   0.003

Velocity (m/s)

Figure 3 CFD analyses of fossil echinoderms. A: Two-dimensional plot (horizontal cross-section) of flow velocity with vectors from simulation of

Dynowski and colleagues (2016) examined flow patterns for the Triassic stalked crinoid *Encrinus liliiformis*. *Encrinus* had ten relatively short and stiff arms, which are thought to have been less flexible than extant forms, necessitating a different feeding posture (Ausich et al., 1999). The functionality of this posture was investigated with CFD. First, digital models with varied arm configurations were constructed. These were then used in computer simulations of inlet velocities of 0.03, 0.14 and 0.5 m/s with models at different orientations to flow, thereby approximating a shallow marine environment with rapidly changing current velocities and directions, as inferred for *Encrinus* (Hagdorn, 1999). Lagrangian particle tracking was used to simulate the trajectories of sub-millimetre particles suspended in the flow, which were taken as representative of planktonic organisms. For validation, flow around a physical model of the base geometry was evaluated in a recirculating flow tank using particle image velocimetry. Comparison of CFD and flume experiments revealed only small differences in the results, establishing the accuracy of the computer simulations. Computational fluid dynamics results for simulations of flow from the aboral side showed a zone of recirculation developed downcurrent of the models, which transported particles back into the crown (Figure 3B). In contrast, simulations of flow from the oral and lateral sides did not create recirculation, but instead resulted in direct transportation of particles into the crown. Thus, the feeding posture of *Encrinus liliiformis* ensured it was able to feed effectively in environments characterized by frequently changing flow conditions.

Caption for Figure 3 (cont.)

water flow around a model of the cinctan *Protocinctus mansillaensis* orientated with the mouth downcurrent, assuming active feeding. Direction of ambient flow from left to right. Modified from Rahman and colleagues (2015, fig. 2i), reproduced with permission from the Royal Society. B: Three-dimensional plot with path lines from simulation of water flow around a model of the crinoid *Encrinus liliiformis* orientated with the stalk and crown bent downcurrent. Direction of ambient flow from left to right. Modified from Dynowski and colleagues (2016, fig. 10A), reproduced under a CC BY 4.0 license. C: Three-dimensional plot of flow velocity within the hydrospire from a simulation of water flow around a model of the blastoid *Monoschizoblastus rofei* orientated with the stem and theca bent downcurrent, assuming active flow. Direction of ambient flow from left to right. Modified from Waters and colleagues (2017, fig. 5.5), reproduced with permission from Cambridge University Press.

Waters and colleagues (2017) analysed the function of blastoid hydrospires. Previous studies of fluid flow through these structures proposed that they were effective gas exchange surfaces (Schmidtling & Marshall, 2010; Huynh et al. 2015), but this work assumed the hydrospires were orientated vertically in life; instead, a posture with the hydrospires orientated horizontally has been suggested to be the most common feeding strategy in blastoids (Breimer & Macurda, 1972). Thus, CFD was used to evaluate hydrospire function in this alternative reconstruction of blastoid living position. A digital model of the blastoid *Monoschizoblastus rofei* was produced, with a legacy data set of serial acetate peels used to reconstruct the internal anatomy. Water flows with inlet velocities of 0.005, 0.02, 0.05 and 0.1 m/s were simulated around the model, which was orientated with the stem and theca bent downcurrent and the hydrospires horizontal. Both passive and active flow through the hydrospires were simulated by assigning different exit velocities to the excurrent orifice (spiracle). The results showed that the simulations of passive flow resulted in significant respiratory leakage, with most water exiting through the adoralmost hydrospire pores, rather than the spiracle. However, in the simulations of active flow with excurrent velocities around 50 per cent of the inlet velocity, flow largely exited through the spiracle with little leakage in other parts of the hydrospires (Figure 3C). This suggests that active cilia-driven water flow through the hydrospires was important for efficient gas exchange, particularly in environments with variable current velocities.

5 Emerging Applications and Future Directions

As noted earlier, CFD is not yet widely used in echinoderm paleobiology. The case studies outlined in the previous section exemplify the types of questions that can be addressed with this approach (see Rahman, 2017 for further examples), demonstrating its potential utility. Moreover, the availability of increasingly powerful computer hardware and software, together with the growing number of virtual models of fossil echinoderms, provides a strong foundation for conducting functional analysis using CFD. However, a solid understanding of fluid mechanics is essential for those wishing to undertake this work, and close collaboration with experts in the field will be beneficial for ensuring the accuracy of analyses.

Replicating and extending existing experimental studies represents a good starting point for paleontological CFD analyses. Where there is close agreement between the results of laboratory experiments and CFD, this provides greater confidence in the accuracy of the computer simulations. Additionally, because computer simulations are typically cheaper, faster and more flexible than experiments in flume tanks, they can be used to expand such studies to include a wider variety of flow parameters and model orientations. For example, Waters and colleagues (2017) established the validity of their computational analyses by digitally recreating the experiment performed by Huynh and colleagues (2015). They then extended this work by carrying out simulations for a full model of the animal in an alternative living posture, using a range of inlet velocities and different assumptions regarding passive and active flow rates. Drawing inspiration from laboratory experiments in this manner will help demonstrate the accuracy of CFD in paleontology, while at the same time enhancing understanding of how extinct echinoderms interacted with moving fluids.

Computational fluid dynamics also offers great potential for testing functional hypotheses in fossil taxa. Suspension feeding is especially amenable to this approach because it relies on flow to feeding structures; the effectiveness of hypothesized feeding strategies in varied environmental conditions can be evaluated by visualizing flow patterns for simulations of different inlet velocities and model orientations (e.g. Shiino et al., 2009; Rahman et al., 2015a, 2015b; Dynowski et al., 2016; Darroch et al., 2017; Gibson et al., 2019). Alternatively, hypotheses regarding the stability of an organism on the seafloor could be tested by quantifying drag and lift forces experienced by models in different positions (e.g. Rahman et al., 2015b; Darroch et al., 2017; Gibson et al., 2019). This hypothesis testing framework can be extended further through comparative analyses of multiple taxa, both within and among groups, thereby enabling the assessment of evolutionary patterns (e.g. Wroe et al., 2018; Gutarra et al., 2019; Troelsen et al., 2019; Hebdon et al., 2020; Rahman et al., 2020). Furthermore, models of hypothetical morphologies could be created to investigate the function of specific characters (e.g. Shiino & Kuwazuru, 2010; Shiino et al., 2012; Rahman et al., 2015a; Darroch et al., 2017; Gutarra et al., 2019; Hebdon et al., 2020). Thus, previously untestable hypotheses are now open to interrogation using CFD, facilitating more rigorous studies of the relationship between form, function and evolution in fossil echinoderms.

Particle tracking is another area of considerable promise. This entails simulating the motion of small particles within the flow, broadly equivalent to determining the trajectories of seeding particles in experimental fluid mechanics. The technique provides a more accurate representation of the path of individual particles than can be obtained with plots of flow vectors and streamlines, and can be used to trace the movement of, for example, planktonic organisms suspended in water (e.g. Dynowski et al., 2016). Virtually tracking particles in this way could be particularly useful for investigating mechanisms

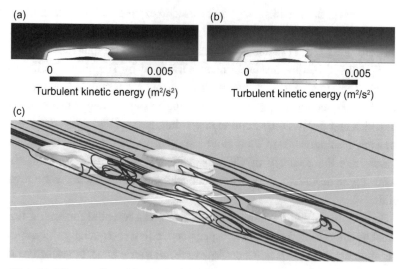

Figure 4 Future directions in CFD analyses. A, B: Two-dimensional plots (vertical cross-sections) of turbulent kinetic energy from simulations of water flow around a model of the cinctan *Protocinctus mansillaensis*, with equivalent sand roughness heights of 0 (A) and 0.01 m (B) applied to the lower surface of the domain. Direction of ambient flow from left to right. C: Three-dimensional plot with streamlines from simulation of water flow around five closely spaced models of the cinctan *Protocinctus mansillaensis*. Direction of ambient flow from top left to bottom right.

of particle capture involved in suspension feeding, or for exploring the dispersal of gametes and larvae.

Thus far, studies applying CFD to fossil echinoderms have used a smooth flat boundary to represent the sediment surface, but this does not accurately reflect the morphology of the seafloor. Bottom roughness, for instance, varies depending on the nature of the substrate, and will influence flow close to the sediment–water interface (Souza & Friedrichs, 2005). Future work could address this by incorporating values of surface roughness into the no-slip boundary condition assigned to the lower surface of the domain (e.g. Figures 4A, 4B). Alternatively, seafloor morphology could be digitally modelled by using a textured surface as the domain floor. This would allow for examination of how changes in substrate type (e.g. those linked to the Cambrian substrate revolution; Bottjer et al., 2000; Dornbos, 2006) influenced fluid flow within the boundary layer, providing insight into the function and ecology of extinct species.

Last, with increasing computer power, CFD simulations incorporating multiple models are becoming more feasible, opening up the possibility to investigate the paleoecology of populations and communities (e.g. Gibson et al., 2019). For example, the possible advantages of aggregation (Lefebvre, 2007) could be evaluated through computer simulations of flow around closely spaced models (e.g. Figure 4C). In addition, the importance of tiering (Bottjer & Ausich, 1986; Dornbos, 2008) could be assessed using models reaching different heights above the floor of the domain. While computationally challenging, extending paleontological CFD analyses to encompass virtual populations and communities will be key for uncovering the complexity of past ecosystems.

6 Conclusions

Echinoderm paleontologists have always been interested in understanding how ancient animals were adapted to life in moving fluids. Historically, experiments in flume tanks were used to study the hydrodynamics of fossil forms. However, the recent growth of CFD in paleontology provides new opportunities for computational functional analysis. This approach is enabling rigorous tests of long-standing hypotheses in extinct echinoderms with important ecological and evolutionary implications. Future improvements in computer power will allow us to tackle increasingly complex questions, shedding light on the paleobiology not only of individual taxa, but also of entire communities and ecosystems.

References

Alexander, D. E. & Ghiold, J. (1980). The functional significance of the lunules in the sand dollar, *Mellita quinquiesperforata*. *The Biological Bulletin*, **159** (3), 561–70.

Álvaro, J. J. & Vennin, E. (1997). Episodic development of Cambrian eocrinoid-sponge meadows in the Iberian Chains (NE Spain). *Facies*, **37**(1), 49–63.

Ausich, W. I., Brett, C. E., Hess, H. & Simms, M. J. (1999). Crinoid form and function. In H. Hess, W. I. Ausich, C. E. Brett & M. J. Simms, eds., *Fossil Crinoids*. Cambridge: Cambridge University Press, pp. 3–30.

Bauer, J. E., Waters, J. A. & Sumrall, C. D. (2019). Redescription of *Macurdablastus* and redefinition of Eublastoidea as a clade of Blastoidea (Echinodermata). *Palaeontology*, **62**(6), 1003–13.

Baumiller, T. K. (1990). Physical modeling of the batocrinid anal tube: Functional analysis and multiple hypothesis testing. *Lethaia*, **23**(4), 399–408.

Baumiller, T. K. (2008). Crinoid ecological morphology. *Annual Review of Earth and Planetary Science*, **36**, 221–49.

Baumiller, T. K., LaBarbera, M. & Woodley, J. D. (1991). Ecology and functional morphology of the isocrinid *Cenocrinus asterius* (Linnaeus) (Echinodermata: Crinoidea): In situ and laboratory experiments and observations. *Bulletin of Marine Science*, **48**(3), 731–48.

Baumiller, T. K. & Plotnick, R. E. (1989). Rotational stability in stalked crinoids and the function of wing plates in *Pterotocrinus depressus*. *Lethaia*, **22**(3), 317–26.

Bottjer, D. J. & Ausich, W. I. (1986). Phanerozoic development of tiering in soft substrata suspension-feeding communities. *Paleobiology*, **12**(4), 400–20.

Bottjer, D. J., Hagadorn, J. W. & Dornbos, S. Q. (2000). The Cambrian substrate revolution. *GSA Today*, **10**(9), 1–7.

Bourke, J. M., Porter, W. R., Ridgely, R. C., Lyson, T. R., Schachner, E. R., Bell, P. R. & Witmer, L. M. (2014). Breathing life into dinosaurs: Tackling challenges of soft-tissue restoration and nasal airflow in extinct species. *The Anatomical Record*, **297**(11), 2148–86.

Bourke, J. M., Porter, W. R. & Witmer, L. M. (2018). Convoluted nasal passages function as efficient heat exchangers in ankylosaurs (Dinosauria: Ornithischia: Thyreophora). *PLoS ONE*, **13**(12), e0207381.

Breimer, A. & Macurda, D. B., Jr. (1972). The phylogeny of the fissiculate blastoids. *Verhandelingen der Koninklijke Nederlandse Akademie van Wetenschappen, Afdeling Natuurkunde. Eerste Reeks*, **26**(3), 1–390.

Briggs, D. E. G., Siveter, D. J., Siveter, D. J., Sutton, M. D. & Rahman, I. A. (2017). An edrioasteroid from the Silurian Herefordshire Lagerstätte of England reveals the nature of the water vascular system in an extinct echinoderm. *Proceedings of the Royal Society B*, **284**(1862), 20171189.

Clark, E. G., Bhullar, B.-A. S., Darroch, S. A. F. & Briggs, D. E. G. (2017). Water vascular system architecture in an Ordovician ophiuroid. *Biology Letters*, **13**(12), 20170635.

Cohen-Rengifo, M., Agüera, A., Detrain, C., Bouma, T. J., Dubois, P. & Flammang, P. (2018). Biomechanics and behaviour in the sea urchin *Paracentrotus lividus* (Lamarck, 1816) when facing gradually increasing water flows. *Journal of Experimental Marine Biology and Ecology*, **506**, 61–71.

Daley, P. E. J. (1996). The first solute which is attached as an adult: A Mid-Cambrian fossil from Utah with echinoderm and chordate affinities. *Zoological Journal of the Linnean Society*, **117**(4), 405–40.

Darroch, S. A. F., Rahman, I. A., Gibson, B., Racicot, R. A. & Laflamme, M. (2017). Inference of facultative mobility in the enigmatic Ediacaran organism *Parvancorina*. *Biology Letters*, **13**(5), 20170033.

David, B., Lefebvre, B., Mooi, R. & Parsley, R. (2000). Are homalozoans echinoderms? An answer from the extraxial-axial theory. *Paleobiology*, **26**(4), 529–55.

Dec, M. (2019). Hydrodynamic performance of psammosteids: New insights from computational fluid dynamics simulations. *Acta Palaeontologica Polonica*, **64**(4), 679–84.

Dornbos, S. Q. (2006). Evolutionary palaeoecology of early epifaunal echinoderms: Response to increasing bioturbation levels during the Cambrian radiation. *Palaeogeography, Palaeoclimatology, Palaeoecology*, **237**(2–4), 225–39.

Dornbos, S. Q. (2008). Tiering history of early epifaunal suspension-feeding echinoderms. In W. I. Ausich & G. D. Webster, eds., *Echinoderm Paleobiology*. Bloomington: Indiana University Press, pp. 133–43.

Dynowski, J. F., Nebelsick, J. H., Klein, A. & Roth-Nebelsick, A. (2016). Computational fluid dynamics analysis of the fossil crinoid *Encrinus liliiformis* (Echinodermata: Crinoidea). *PLoS ONE*, **11**(5), e0156408.

Emelyanov, E. M. (2005). *The Barrier Zones in the Ocean*. New York: Springer.

Evans, M. W. & Harlow, F. H. (1957). The particle-in-cell method for hydrodynamic calculations. *Los Alamos Scientific Laboratory Report*, **LA-2139**, 1–76.

Friedrich, W.-P. (1993). Systematik und Funktionsmorphologie mittelkambrischer Cincta (Carpoidea, Echinodermata). *Beringeria*, **7**, 3–190.

Gibson, B. M., Rahman, I. A., Maloney, K. M., Racicot, R. A., Mocke, H., Laflamme, M. & Darroch, S. A. F. (2019). Gregarious suspension feeding in a modular Ediacaran organism. *Science Advances*, **5**(6), eaaw0260.

Gutarra, S., Moon, B. C., Rahman, I. A., Palmer, C., Lautenschlager, S., Brimacombe, A. J. & Benton, M. J. (2019). Effects of body plan evolution on the hydrodynamic drag and energy requirements of swimming in ichthyosaurs. *Proceedings of the Royal Society B*, **286**(1898), 20182786.

Hagdorn, H. (1999). Triassic Muschelkalk of Central Europe. In H. Hess, W. I. Ausich, C. E. Brett & M. J. Simms, eds., *Fossil Crinoids*. Cambridge: Cambridge University Press, pp. 164–176.

Harlow, F. H. & Welch, J. E. (1965). Numerical calculation of time-dependent viscous incompressible flow of fluid with free surface. *The Physics of Fluids*, **8**(12), 2182–9.

Hebdon, N., Ritterbush, K. A. & Choi, Y. (2020). Computational fluid dynamics modeling of fossil ammonoid shells. *Palaeontologia Electronica*, **23**(1), a21.

Hess, J. L. & Smith, A. M. O. (1967). Calculation of potential flow around arbitrary bodies. *Progress in Aerospace Sciences*, **8**, 1–138.

Holtz, E. H. & MacDonald, B. A. (2009). Feeding behaviour of the sea cucumber *Cucumaria frondosa* (Echinodermata: Holothuroidea) in the laboratory and the field: Relationships between tentacle insertion rate, flow speed, and ingestion. *Marine Biology*, **156**(7), 1389–98.

Huynh, T. L., Evangelista, D. & Marshall, C. R. (2015). Visualizing the fluid flow through the complex skeletonized respiratory structures of a blastoid echinoderm. *Palaeontologia Electronica*, **18**(1), 14A.

Kogan, I., Pacholak, S., Licht, N., Schneider, J. W., Brücker, C. & Brandt, S. (2015). The invisible fish: Hydrodynamic constraints for predator–prey interaction in fossil fish *Saurichthys* compared to recent actinopterygians. *Biology Open*, **4**, 1715–26.

Lautenschlager, S. (2016). Reconstructing the past: Methods and techniques for the digital restoration of fossils. *Royal Society Open Science*, **3**(10), 160342.

Lefebvre, B. (2007). Early Palaeozoic palaeobiogeography and palaeoecology of stylophoran echinoderms. *Palaeogeography, Palaeoclimatology, Palaeoecology*, **245**(1–2), 156–99.

Liu, S., Smith, A. S., Gu, Y., Tan, J., Liu, K. & Turk, G. (2015). Computer simulations imply forelimb-dominated underwater flight in plesiosaurs. *PLoS Computational Biology*, **11**(12), e1004605.

Loo, L.-O., Jonsson, P. R., Sköld, M. & Karlsson, Ö. (1996). Passive suspension feeding in *Amphiura filiformis* (Echinodermata: Ophiuroidea): Feeding behaviour in flume flow and potential feeding rate of field populations. *Marine Ecology Progress Series*, **139**, 143–55.

Macurda, D. B., Jr. & Meyer, D. L. (1974). Feeding posture of modern stalked crinoids. *Nature*, **247**(5440), 394–6.

Messing, C. G., RoseSmyth, M. C., Mailer, S. R. & Miller, J. E. (1988). Relocation movement in a stalked crinoid (Echinodermata). *Bulletin of Marine Science*, **42**(3), 480–7.

Parsley, R. L. (1990). *Aristocystites*, a recumbent diploporid (Echinodermata) from the Middle and Late Ordovician of Bohemia, ČSSR. *Journal of Paleontology*, **64**(2), 278–93.

Parsley, R. L. (1999). The Cincta (Homostelea) as blastozoans. In M. D. Candia Carnevali & F. Bonasoro, eds., *Echinoderm Research 1998*. Rotterdam: Balkema, pp. 369–75.

Parsley, R. L. (2015). Flume studies using 1:1 scale models of Series 2 and basal Series 3 Cambrian gogiid eocrinoids from Guizhou Province, China to determine feeding posture and mode of attachment. *Palaeoworld*, **24**(4), 400–7.

Rahman, I. A. (2017). Computational fluid dynamics as a tool for testing functional and ecological hypotheses in fossil taxa. *Palaeontology*, **60**(4), 451–9.

Rahman, I. A., Darroch, S. A. F., Racicot, R. A. & Laflamme, M. (2015a). Suspension feeding in the enigmatic Ediacaran organism *Tribrachidium* demonstrates complexity of Neoproterozoic ecosystems. *Science Advances*, **1**(10), e1500800.

Rahman, I. A. & Lautenschlager, S. (2017). Applications of three-dimensional box modeling to paleontological functional analysis. In L. Tapanila & I. A. Rahman, eds., *Virtual Paleontology: The Paleontological Society Papers*, **22**, 119–32.

Rahman, I. A., O'Shea, J., Lautenschlager, S. & Zamora, S. (2020). Potential evolutionary trade-off between feeding and stability in Cambrian cinctan echinoderms. *Palaeontology*, 63(5), 689–701.

Rahman, I. A., Thompson, J. R., Briggs, D. E. G., Siveter, D. J., Siveter, D. J. & Sutton, M. D. (2019). A new ophiocistioid with soft-tissue preservation from the Silurian Herefordshire Lagerstätte, and the evolution of the holothurian body plan. *Proceedings of the Royal Society B*, **286**(1900), 20182792.

Rahman, I. A. & Zamora, S. (2009). The oldest cinctan carpoid (stem-group Echinodermata), and the evolution of the water vascular system. *Zoological Journal of the Linnean Society*, **157**(2), 420–32.

Rahman, I. A., Zamora, S., Falkingham, P. L. & Phillips, J. C. (2015b). Cambrian cinctan echinoderms shed light on feeding in the ancestral deuterostome. *Proceedings of the Royal Society B*, **282**(1818), 20151964.

Reich, M., Sprinkle, J., Lefebvre, B., Rössner, G. E. & Zamora, S. (2017). The first Ordovician cyclocystoid (Echinodermata) from Gondwana and its morphology, paleoecology, taphonomy, and paleogeography. *Journal of Paleontology*, **91**(4), 735–54.

Riddle, S. (1989). Functional morphology and paleoecological implications of the platycrinitid column (Echinodermata, Crinoidea). *Journal of Paleontology*, **63**(6), 889–97.

Rigby, S. & Tabor, G. (2006). The use of computational fluid dynamics in reconstructing the hydrodynamic properties of graptolites. *GFF*, **128**(2), 189–94.

Saulsbury, J. & Zamora, S. (2019). The nervous and circulatory systems of a Cretaceous crinoid: Preservation, palaeobiology and evolutionary significance. *Palaeontology*, **63**(2), 243–53.

Schmidtling, R. C., II, & Marshall, C. R. (2010). Three dimensional structure and fluid flow through the hydrospires of the blastoid echinoderm, *Pentremites rusticus*. *Journal of Paleontology*, **84**(1), 109–17.

Shiino, Y. & Kuwazuru, O. (2010). Functional adaptation of spiriferide brachiopod morphology. *Journal of Evolutionary Biology*, **23**(7), 1547–57.

Shiino, Y. & Kuwazuru, O. (2011). Comparative experimental and simulation study on passive feeding flow generation in *Cyrtospirifer*. *Memoirs of the Association of Australasian Palaeontologists*, **41**, 1–8.

Shiino, Y., Kuwazuru, O., Suzuki, Y. & Ono, S. (2012). Swimming capability of the remopleuridid trilobite *Hypodicranotus striatus*: Hydrodynamic functions of the exoskeleton and the long, forked hypostome. *Journal of Theoretical Biology*, **300**, 29–38.

Shiino, Y., Kuwazuru, O., Suzuki, Y., Ono, S. & Masuda, C. (2014). Pelagic or benthic? Mode of life of the remopleuridid trilobite *Hypodicranotus striatulus*. *Bulletin of Geosciences*, **89**(2), 207–18.

Shiino, Y., Kuwazuru, O. & Yoshikawa, N. (2009). Computational fluid dynamics simulations on a Devonian spiriferid *Paraspirifer bownockeri* (Brachiopoda): Generating mechanism of passive feeding flows. *Journal of Theoretical Biology*, **259**(1), 132–41.

Siedler, G., Griffies, S. M., Gould, J. & Church, J. A., eds. (2013). *Ocean Circulation and Climate: A 21st Century Perspective*. Oxford: Academic Press.

Smith, A. B. (2005). The pre-radial history of echinoderms. *Geological Journal*, **40**(3), 255–80.

Souza, A. & Friedrichs, C. (2005). Near-bottom boundary layers. In H. Z. Baumert, J. Simpson & J. Sündermann, eds., *Marine Turbulence: Theories, Observations, and Models*. Cambridge: Cambridge University Press, pp. 283–296.

Stow, D. A. V., Hernández-Molina, F. J., Llave, E., Sayago-Gil, M., Díaz del Río, V. & Branson, A. (2009). Bedform-velocity matrix: The estimation of bottom current velocity from bedform observations. *Geology*, **37**(4), 327–30.

Telford, M. (1983). An experimental analysis of lunule function in the sand dollar *Mellita quinquiesperforata*. *Marine Biology*, **76**(2), 125–34.

Thompson, M., Drolet, D. & Himmelman, J. H. (2005). Localization of infaunal prey by the sea star *Leptasterias polaris*. *Marine Biology*, **146**(5), 887–94.

Troelsen, P. V., Wilkinson, D. M., Seddighi, M., Allanson, D. R. & Falkingham, P. L. (2019). Functional morphology and hydrodynamics of plesiosaur necks: Does size matter? *Journal of Vertebrate Paleontology*, **39** (2), e1594850.

Waters, J. A., Sumrall, C. D., White, L. E., & Nguyen, B. K. (2015). Advancing phylogenetic inference in the Blastoidea (Echinodermata): Virtual 3D reconstructions of the internal anatomy. In S. Zamora & I. Rábano, eds., *Progress in Echinoderm Palaeobiology. Cuadernos del Museo Geominero*, 19, 193–7.

Waters, J. A., White, L. E., Sumrall, C. D. & Nguyen, B. K. (2017). A new model of respiration in blastoid (Echinodermata) hydrospires based on computational fluid dynamic simulations of virtual 3D models. *Journal of Paleontology*, **91**(4), 662–71.

Welch, J. R. (1978). Flume study of simulated feeding and hydrodynamics of a Paleozoic stalked crinoid. *Paleobiology*, **4**(1), 89–95.

Wroe, S., Parr, W. C. H., Ledogar, J. A., Bourke, J., Evans, S. P., Fiorenza, L., Benazzi, S., Hublin, J.-J., Stringer, C., Kullmer, O., Curry, M., Rae, T. C. & Yokley, T. R. (2018). Computer simulations show that Neanderthal facial morphology represents adaptation to cold and high energy demands, but not heavy biting. *Proceedings of the Royal Society B*, **282**(1876), 20180085.

Zamora, S., Rahman, I. A. & Smith, A. B. (2012). Plated Cambrian bilaterians reveal the earliest stages of echinoderm evolution. *PLoS ONE*, 7(6), e38296.

Zamora, S. & Smith, A. B. (2008). A new Middle Cambrian stem-group echinoderm from Spain: Palaeobiological implications of a highly asymmetric cinctan. *Acta Palaeontologica Polonica*, **53**(2), 207–20.

Zamora, S. & Smith, A. B. (2012). Cambrian stalked echinoderms show unexpected plasticity of arm construction. *Proceedings of the Royal Society B*, **279**(1727), 293–8.

Acknowledgements

I thank Colin Sumrall for the invitation to submit this Element. I am grateful to Brad Deline, Brandt Gibson, Johnny Waters and Tom Baumilller for comments on earlier versions of the text. I was funded by Oxford University Museum of Natural History.

Cambridge Elements ☰

Elements of Paleontology

Editor-in-Chief

Colin D. Sumrall
University of Tennessee

About the Series

The Elements of Paleontology series is a publishing collaboration between the Paleontological Society and Cambridge University Press. The series covers the full spectrum of topics in paleontology and paleobiology, and related topics in the Earth and life sciences of interest to students and researchers of paleontology.

The Paleontological Society is an international nonprofit organization devoted exclusively to the science of paleontology: invertebrate and vertebrate paleontology, micropaleontology, and paleobotany. The Society's mission is to advance the study of the fossil record through scientific research, education, and advocacy. Its vision is to be a leading global advocate for understanding life's history and evolution. The Society has several membership categories, including regular, amateur/avocational, student, and retired. Members, representing some 40 countries, include professional paleontologists, academicians, science editors, Earth science teachers, museum specialists, undergraduate and graduate students, postdoctoral scholars, and amateur/avocational paleontologists.

Cambridge Elements ☰

Elements of Paleontology

Printed in the United States
By Bookmasters